Friends of Tinkle Rescue Club

A true story of
TINKLE the Siamese
and ANN her owner

By Ann Eve Moss

Illustrated by Henry S. Gillette
Edited by Alwyn Moss

Pocahontas Press, Inc.
Blacksburg, Virginia

Friends of Tinkle Rescue Club

Typesetting by Faulkner Printing, Blacksburg, Virginia.
Printed and bound in the United States of America by
Commonwealth Press, Inc., Radford, Virginia.

Pocahontas Press, 2805 Wellesley Court, Blacksburg, Virginia 24060-4126.

First Printing, 1991.

ISBN 0-936015-27-6 Friends of Tinkle Rescue Club

Library of Congress Cataloging In Publication Data:

Moss, Ann Eve, 1903-1989.

Friends of Tinkle Rescue Club : a true story of Tinkle the Siamese and Ann her owner / by Ann Eve Moss ; illustrated by Henry S. Gillette ; edited by Alwyn Moss.

p. cm.

ISBN 0-936015-27-6 (paper)

1. Cat owners—Psychology. 2. Tinkle (Cat) 3. Human-animal relationships. 4. Moss, Ann Eve, 1903-1989. 5. Cat owners—Biography. I. Moss, Alwyn. II. Title.

SF445.5.M68 1991

636.8'25—dc20

90-24192

CIP

Preface

In these years of rapid change, of transience of culture and place, of loss and uncertainty, this story of enduring love between an aging woman and her old-young cat has meaning for many of us, no matter what our age, but is of particular appeal to the young and those growing older as it tells of the mysteries of age and youth in a charming and original way.

Written by a woman nearly eighty who kept a childlike joy in life until her death at 84, this book tells about a relationship Ann had with an animal who was also unique. Small and kittenlike as she was, the Siamese cat "Tinkle" was, by human standards, almost ancient; over twenty-three years old, which, as humans reckon years in cat-life, is about one hundred forty years. And for all this time Tinkle and Ann shared their lives together. As Ann observes, "She's kept me young."

Very timely too in its delicate recording of how an interspecies relationship can give meaning and added richness to both human and non-human life, this little book—published as a memorial to its author, Ann Moss—a writer, painter, and in her early years dancer, singer, and Zeigfeld Follies' Girl—has much to offer everyone. The FTRC (Friends of Tinkle Rescue Club) is also a tribute to the loyalty and good will of those friends.

Alwyn Moss

Tinkle, the small Siamese cat, her two slim front legs straight together on the white balcony ledge, looked like a pale Egyptian statue as she faced the blue spring sea sitting upright, very quiet and seemingly serene. In fact, she was troubled and unhappy, with those curious feelings again, as if something bad was going to happen—*soon*. They were the sort of feelings she had had before—whenever a big disturbing change in her life was coming—and she *knew*.

The trouble this time was hard to understand.

This time the trouble was that Ann, Tinkle's owner, her own special person since her early kittenhood—had to go away. She had to leave their snug little apartment that looked over the turquoise Mediterranean Sea on the coast of southern France to go live with Mary-Lynn, her daughter in England who was soon going to have a baby and who would need her, Ann said, for quite a while—perhaps years.

But Ann had to go there *without Tinkle.* Tinkle could not! Because—and this was what was so hard to understand—in England there is a "law" that all animals entering that country have to be put into something they call "QUARANTINE". Must? Why?—Tinkle seemed to be asking Ann, who replied: "To make certain that visiting animals have no diseases—like *rabies,* a very dangerous disease for animals that might spread to English cats and dogs—and even to people."

So it was a good law.

But QUARANTINE meant to stay in a cage for months and months!

"Imprisoned for months in Quarantine!" Tinkle and Ann moaned together in dismay.

Certainly this was a crisis. The most serious one that Ann and Tinkle had ever faced in their long years together—years of sharing—sharing everything: meals, games, bedtime and sleeping, walks, music, warm evenings near the heater in winter on the soft blue velvet couch, listening to the stereo. Tinkle was very fond of the Big B's, especially Bach and Beethoven, just as Ann was.

And when Ann was busy typing—letters or books or articles, for she was a writer, Tinkle collaborated by catching a letter key from time to time with her paw before it struck the typewriter ribbon, and if Ann was sketching, something she loved to do, Tinkle could trace Ann's spidery pencil lines on the paper with her paw—very gently. Some people thought this a surprising thing for a cat to do. But then there are many things most people don't know about some cats.

Even household chores were a joint venture. For Tinkle always tried to "help" Ann when she swept the kitchen floor by playing with the bristles of the broom, or if Ann was carrying garbage to the back door Tinkle knew how to walk *right* in front of her to show the way....

Together they had built up fine routines, ways of knowing what to expect of each other. But now? What could be done? Ann had to go to her daughter to help her with the new child, and then stay on to allow her to go back to her job. Tinkle could not come. And so far, at any rate, there was no sign of a suitable home for Tinkle, one where she might be happy.

Ann turned to friends to talk about her problem.

Retired, after many years as an assistant editor for a magazine in New York City, Ann had decided to go to France to live—partly to be in the same part of the world, or at least on the same side of the, ocean as her only daughter who had married an Englishman; partly because she had learned French early in life and loved the country; and partly because an apartment on the southern coast of France seemed as if it would be a good place to write books and paint.

And she knew a few people who had settled there.

Eva, for instance, a German-born nurse, had been a neighbor and friend in the States before moving to France. She too loved animals, but had a preference for birds. Ding and Bell, two big bright parrots, lived in her tiny kitchen, and she kept a few smaller birds in her bedroom.

Pearl and Tom, retired high-school teachers from Milwaukee, whom Ann had met, soon after arriving, at a church bazaar. They kept a big fat Bassett hound who waddled duck-like around their garden.

Elizabeth was the person that Ann often felt she shared most with: a painter who supported herself by illustrating children's books. Elizabeth, Swedish by descent, was tall, fair of hair and skin, sensitive but always cheerful, and a "cat person".

Her cat, Topolino, a long-haired silvery tiger-cat on whom she lavished attention (as Ann did on Tinkle) was Tinkle's only real cat friend, though they had met only once or twice.

Then there was Marjorie, who played the piano and violin and who seemed more than pleased to teach Ann (and Tinkle) about music—even bringing tapes to hear or borrow.

Yet despite these fine friends and others, for one reason or another so far there was no sign of a possible home for Tinkle, one where the little cat might feel safe and at peace after Ann set off for England.

And time was passing. Wrapping dishes in paper, settling them in boxes, washing down shelves, Ann tried to feel cheerful while Tinkle watched her from her favorite windowledge, as Ann then began to chop vegetables for dinner.

Yes, happiness for Tinkle were these peaceful hours they spent together, even if Ann was busy baking bread or watering plants on the sunny terrace, as long as she could be nearby. Even waiting for her when she set off mornings for the market with the little string shopping bag (the customary way to shop in France), patiently amusing herself as best she could—like wiggling herself into a small wastebasket, making herself into a round furry ball, then leaping out, always surprising Ann when she came in.

And Ann always laughed!

Not that she was perfect. Oh yes, they too had their ups and downs, she and Ann. Sometimes there was a disagreement about food, for instance. Tinkle had come to live with Ann just after she had been weaned from her Siamese mother—Ann had never had a cat before (though she had two children); but though she was nearly reaching her 70th birthday—like a new and nervous mother, she bought and read as many "How to raise a Kitten into a Healthy Cat" sort of books as she could.

One of the books said that chicken livers were specially good for a growing cat—so Tinkle soon became used to eating chicken liver—more and more, until soon she would eat nothing but that—since it was almost all she had ever eaten.

Years later, in France, Ann read another book that a friend said was the newest and best book on the subject—and the latest thinking—which was quite different from the all-chicken-liver theory....

This book said that cats needed all sorts of variety in their diet, so one day Tinkle found herself face to face with a strange mixture of things in her blue bowl—goat milk, bread, cottage cheese, egg, and some other un-recognizable stuff....

"Come and eat, Tinkle. This is 'Tinkle's casserole'."

For a while Tinkle just sat and stared at this unusual cat-fare. Then she walked away. At dinner-time Ann put it down again—this time she had added grated carrot, minced lettuce (after all, she reasoned, all cats like to eat grass and most vets say they need it), some buttered vegetables, and even potatoes—Carrots? Potatoes? Tinkle would not even sniff at this mess....

It became a battle. Their first. Tinkle turned away every time. By evening Ann gave in, and put something more conventional in Tinkle's blue bowl—chicken liver, milk, some dried cat food. But she did so reluctantly, and not sure what else to do....

She was extremely grateful, then, when she ran into Eva at the city market one morning. Eva was filling her string bag with onions, Brussels sprouts, and kale, and, with a clear and fixed eye, Eva, lifting a finger, said: "Ann. Be firm!—No liver!"

Tinkle picked up the message from the way Ann walked in the door, looked at her—waiting near her feeding place—went to the 'fridge and put down the "Casserole". Still she would have none of it—until, one afternoon, soon after, particularly hungry, she decided to taste this weird thing Ann was so keen to have her eat. MMMM...it wasn't so bad. She 'specially liked the taste of the goat milk and egg.

Soon Tinkle was waiting eagerly for her afternoon casserole—even when there might be some extra surprise in it like corn or sweet potato (which some cats adore), and Ann took pleasure in watching Tinkle, with her little cat grin and pink tongue darting all around her mouth, licking up every smidgin of this "afternoon casserole"...and then washing herself carefully after.

This morning Ann was too busy to pay much attention to Tinkle. Opening, closing drawers, on her knees looking for this blouse, that sweater, for papers she couldn't find—It seemed as if whenever she really needed a particular document or letter or bill she could only remember having put it in some very "special place." But where?

Shocked suddenly by the sound of the too-loud doorbell (that she'd always meant to have softened), Ann jumped up and headed for the door. Tinkle, as usual, ran just ahead of her, then sat directly, neatly, in front of the big wooden door while Ann worked with the stiff brass knob.

"Marjorie! Come in. I didn't know you were back."

"Got back last night. Came right over. Heard about your problem."

Ann led the way to the kitchen, put a kettle of hot water on the stove for tea. Marjorie sat down, took her cotton gloves off, folded them into her large white purse, and set that down on the floor next to her chair.

"How old *is* Tinkle now?" she asked, getting to the point.

Tinkle's ears shifted forwards as she heard her name.

Tinkle liked Marjorie, even if she didn't always pet her or even look at her. Still, Tinkle could tell that Marjorie had a warm heart for animals. As a cat Tinkle found it hard to tell ages, cat or human ages, but Marjorie even looked something like Ann, was about the same size—the height of the refrigerator, walked light, almost on her toes, like Ann, spoke high—but best of all, Tinkle remembered the time that Marjorie had come with her violin and played so beautifully—*Mozart*—while Tinkle and Ann sat listening on the couch....

"Tinkle—is over twenty-two years now." There was a tone of some undisguised pride in Ann's voice.

"Twenty-two years! Ann, are you sure?" Marjorie sounded very skeptical.

"Quite sure," Ann said gently but firmly. "I know the exact date I got her. She was not quite three months old."

"But that's amazing," Marjorie said, turning to study Tinkle who was washing herself after a few nibbles of chicken still in her plate.

"Yes. And she really isn't that much different than when I got her. She still plays, runs, jumps if I throw a ball—or a..."

"So what *will* you do? You could have her put down, you know. She's had a good life..."

"Marjorie!" Ann was shocked to hear such a term coming from her friend.

"Oh, well. I didn't really mean it."

But Tinkle, who had never heard that expression before, just went on carefully washing her smooth pale paw, while waiting for Ann to throw the—her favorite "toy"—a chicken foot, from a *real* chicken that Ann had bought to eat, and which all dried out now was great fun for Tinkle to play with. What a frolic. Tinkle would begin by throwing the foot high in the air from her mouth, straight up in the air. Then she would push it and slide it with her paw toward Ann, until her owner, well-trained by now, would toss it back.

"Oh, Ann. What *is* that thing?" Marjorie asked in some horror.

Ann laughed, just as Tinkle managed to get the treasured object back, and ran up, dropping it at Ann's feet like a trophy—a very special present for her—but then immediately picking it up again ran out of the kitchen with it in her mouth....

"Twenty-two years, did you say?" said Marjorie.

Now, twenty-two years, dear friends, for a cat, is about, they say, the equivalent of 140 years for a human adult. A one-year-old cat is not a baby, as a one-year-old child is, but more like a seven-year-old human. At least that's the way people who study animal life and behavior explain animal age. Cat age in this case. So Tinkle was no ordinary cat, and those who watched her play and eat and run with her quick little body and bright clear eyes were amazed when they learned "how old" she was.

But then some people were amazed to learn "how old" Ann was, for she looked far younger than her nearly seventy years, with her light, easily moving body, thin and graceful carriage, and eyes that could light up like a child's. Her reply was often—"that's because of Tinkle. She's kept me young."

After Marjorie left, Ann sat looking at the yellow kitchen wall. Marjorie couldn't take Tinkle because she travelled so much. Eva couldn't take her, obviously, because of her birds. Betsy's landlord wouldn't allow any pets no matter how big a deposit one offered him. Elizabeth was only allowed "one pet under 10 lbs." (and she had Topie, of course). Tom couldn't because Pearl, his wife, was "allergic to feline dander" (though Tinkle's short smooth fur had little)—and of course they had their dog. There didn't seem to be anyone else to ask. Ann was beginning to lose hope.

"Tinkle, Tinkle," she said, picking the light little lithe body up and tucking her into her aproned lap—"What on earth are we going to do?"

Then Tinkle settled down completely and decided to go to sleep as much to shut out the vibrations of Ann's distress as anything else, while Ann felt a tear gathering in her right eye.

How much she did care for this little creature who seemed to understand so much and who, though wordless, was always communicating with her body and movements and sounds, who was delicate and gentle and, to add to her virtues, cat though she was, had never, as Ann almost boasted to her friends—had never killed anything—"No, she hasn't," she persisted—well at least she had never seen it. Not even a housefly.

To be more exact, Tinkle did catch and play with flies sometimes, her teeth chattering with excitement like most cats—click-buzz, click-buzz. But she never actually caught one or held one for long, so that even after many minutes of this game (for that's what it seemed to be to her), the fly, released, flew off seemingly undamaged in any way at all.

Imagine, Ann marveled, her heart overflowing with love for her gentle cat—"Saint Tinkle!"

Though in her more balanced view of things she realized this might be because Tinkle's Siamese mother, also an indoor city cat, had never taught her kits to hunt or catch or kill things to eat. —"And we are what we learn, aren't we?" Ann commented to herself. "Certainly if we are cats."

"Oh!" said Ann, waking up in the middle of the night. "Marie. Marie!" I forgot about Marie— She loves Tinkle. She might take her—and Tinkle loves her."

Marie lived on the first floor of the same apartment building. Her mother was the caretaker of the house (or *Concierge*, as they call it in France). From the first moment Marie had seen the little Siamese one day, walking on her long red leash, stepping as delicately as a trained pony, Marie had been "enchantée avec la petite chatte" and would come often after school to pay Tinkle a visit and play with her if Ann was busy.

Or watch Ann play with her.

So when Ann called Marie the following afternoon, it was with some new hope that this might be an answer to the ever more imminent and difficult problem of Tinkle's future....

"Entrez, Marie, entrez—"

Marie was nine years old, with silky, shoulder-length, dark brown hair and cherry brown, very round eyes—sometimes she talked so quietly Ann could scarcely hear her—and she spoke only French—which Ann of course understood well after ten years in France.

Tinkle emerged from somewhere and began to rub up against Marie's knee-high blue socks, and Marie bent down to pick her up.

How wonderful it would be, Ann was thinking, *if—*

"Fâites-la parler Madame—Je vous emprie..." half-whispered Marie—which meant, "Make her talk, please..."

It was something that Tinkle did—one of her many mysterious surprises—that Ann too found amusing and delightful—to hear the small cat talking—in unmistakably changing and meaningful tones—

So Ann began—making a crooning noise deep in her throat the way some cat mothers do to their kitten—and almost at once Tinkle, sitting between Marie and Ann, began to croon back in her deep, yet clear Siamese voice. Then Ann made her meow higher, higher—then lower and lower—And Tinkle followed the expressive scale almost perfectly—as if mimicking Ann's sounds, whether soft contented little cries, or long complaining cries, or just exclamations.

"Quelle conversation!" said the little girl, in her usual rapt amazement...and she picked Tinkle up and gave her a hug—while Ann beamed....

Marie, such a careful and loving child, could give Tinkle some of the affection and play she was used to—and the home, which Ann had visited often, was clean and pleasant....

But Marie's mother said "Non." Said that she herself was too busy to take care of "la petite chatte"—that Marie had to *study*, the little cat would distract her—and add to the work of the household.

Alas, another disappointment—

Wednesday mornings had become a special tradition with Ann and her friends. They almost always met together in one of the coffeehouses along the beach for croissants and coffee—or tea—to share events and problems of their lives. But this Wednesday morning Ann felt tired and sad and not inclined to go anywhere. She had told this to Marjorie and to Elizabeth the night before. But at 10 a.m. her doorbell rang—and rang.

There stood Elizabeth in her long blue, belted velvet coat and matching beret, purse over her arm.

"Let's go, Annie..." she said, rather loudly for her.

And no amount of explanations would put her off, so with a few gentle pats and an apology to Tinkle off they went—down the elevator, out the heavy iron-trellis door to Elizabeth's ancient, oversized four-door sedan. Marjorie, Eva, and Betsy in the back seat, and Ann helped into the seat next to Elizabeth, the driver.

"What is all this?" asked Ann once they were settled at one of the round glass tables near the terrace windows looking out on the sea.... She had already suspected this might be a sort of surprise farewell party for her from the almost child-like mischievous expressions on the faces of the four women...who looked at each other expectantly....

"We have an idea," Marjorie began—"Actually it was Elizabeth's idea—of course."

Ann began to feel somewhat bewildered—"An idea?"

"We've always been a sort of club—almost four years or so now—haven't we?" Marjorie continued....

Ann nodded, and waited.

"But now," Betsy took over, "we've decided we need a name. And we've found one." Betsy looked over to Elizabeth, encouraging her to speak.

"We are now—officially," said Elizabeth in her more normal quiet voice, "the—*Friends of Tinkle Rescue Club.*"

Ann looked up and around the table at each one of them.

"It's a nice name, isn't it?" Eva asked Ann.

"And—it will mean that you can stop worrying so much about your little old cat..." said Marjorie.

"She's not *old,*" retorted Ann. "But how very wonderful."—Pause. Then, lifting her shoulders, "But what does it mean?"

"Well, we don't know exactly yet," said Elizabeth, "except that we're committed to helping you and Tinkle work this out—somehow. Even though for all our various reasons none of us can actually give Tinkle a home, much as we might want to—still, we can rally all our forces, and other friends, and.... In other words, Ann, you and Tinkle are not alone in this problem."

A rush of tears, some of which overflowed Ann's lids, and Eva moved over to put her arm around her friend....

Before their morning meeting was over, the newly named club had made some decisions. The first was to put an AD in the local papers—some in French, some in English—for "a good loving home for wonderful Siamese cat—"

"A good loving home?"—Tinkle questioned sadly.

Tinkle knew what it was to be given away because it had happened twice before. Twice? Yes.

Once Ann had had to go away for some work she could only do at a conference center, where no pets were allowed, and Ann decided she would "temporarily" give Tinkle away for a few months. And so she did—leaving a trembling little Tinkle in a strange house with a very nice couple who, however, had very little time or interest to give to cats.... And many miles away as she was, Ann found that she could not sleep very well or work for worrying—

So back she came after just a week or so, racing to the place she'd left her— and, whisking her up in her arms, brought her safely home—

That night, back in their own apartment, something strange occurred. The rescued Tinkle, stretched out quite straight on the bed beside the lying-down body of her mistress, purred. Purred. Purred and purred so loud for so many hours that Ann became alarmed— "Oh dear me. What have I done? Something must be wrong—her purrer is out of order—she can't stop.".... But in the morning when Ann ordered a taxi to rush Tinkle to the vet—The vet told her there was nothing really wrong with Tinkle except that she was so happy to be back with Ann.

And that was true....

With such a memory, Ann found it even harder to think of giving Tinkle away—again.

Elizabeth, Eva, Betsy, Tom—and even Marjorie—assured Ann that they would somehow find a truly good home for Tinkle this time, and that they would all make it a point of paying regular visits to see Tinkle wherever she was. Marie promised too, and her mother. And even the vet who had a very special feeling for Tinkle "because of her amazing *youthful* age." Perhaps, Ann thought, he took a little credit for this phenomenon. Even he promised Ann he would be another guardian angel for "la belle petite chatte"—So Ann went to Tinkle to try to explain how many people cared about her and would make it all right....

Still Tinkle remembered the second time she'd been given away. A terrible winter ordeal no cat could ever forget—

Ann had caught a cold in the month of January. By February it had turned, the doctor said, into pneumonia, and it was necessary for Ann to go to the hospital late one night. Recognizing how sick she was, Ann decided the day before to give Tinkle away—"for good" to a lady who promised to give Tinkle *a very good home*. But, the lady said, it had to be "forever" and she made Ann promise she would never take Tinkle back. Weak and sick, with no way to foresee the future, Ann promised.

A few weeks later, when Ann recovered, she was able to go home. A home without Tinkle. Without her beloved cat-friend. It hurt terribly. Ann was heartsick.

But the lady who now had Tinkle would never give her back to Ann. It was "forever," and Ann, determined to keep her word, wrote only a letter or two to the lady inquiring about Tinkle's health and happiness.

The lady never answered. Then Ann began to worry, not knowing how Tinkle was, really, and feeling responsible for having given her away so quickly. She kept hearing Tinkle's "mrroww" and imagining Tinkle's expressions—

So Ann prayed to know, *in some way*, if Tinkle was indeed satisfied and content in the new home. She even promised herself that "if Tinkle is all right and fairly happy there, I'll accept her being out of my life—especially if she can gladden someone else's life."

Not long after this time Ann had a dream. It seemed to her a strange dream, unlike any she'd ever had before. Could it be—a message from Tinkle?

In this dream, Tinkle was in the lady's kitchen—plates, pots, forks and knives all heaped around her face and mouth, piled up to and right over her little nose so that it looked as if she couldn't breathe—as if she was gasping for breath.

"What's this?" In alarm Ann cried. "What does this dream mean?" For Ann had heard of dreams that told people things that were true, and she believed such things were possible.

So, the very next day Ann ran through the streets to the house where Tinkle lived—hoping to be able to speak with the lady. There was no elevator. The apartment was on the third floor and Ann's heart was beating hard. She rang the pull-bell a third time, and at last the door opened slowly.

A woman she scarcely recognized was standing in the doorway in a bathrobe. She waved Ann inside, in front of her, and followed. Ann looked round for Tinkle but didn't see her. Instead she sat down in a rocker while the lady went into another room.... In a minute she came back—with the little cat in her arms.

But Ann could scarcely believe what she saw. The dream was true. True! A scrawny, miserable-looking creature that seemed near death. Ann stood astonished, afraid even to touch her.

But at the sight of Ann, Tinkle, as if in great relief, gathered herself up and, with one obviously difficult effort costing all her strength, leaped at Ann and landed on her shoulder as if onto her tree of life. And Ann closed her arms around the cat who relaxed visibly there.

Ann looked at Tinkle's new owner questioningly. But the lady, poor lady, was obviously herself not very well. She did not explain much to Ann other than to say that while Ann was in the hospital she had had to change jobs. The new job meant very late hours so she needed to sleep in the daytime and was very tired much of the time—"Très fatiguée, Madame." And this meant that Tinkle had very often been alone, and that the lady had found it very difficult to take care of the house, fix her meals, or even feed Tinkle regularly, who didn't want to eat anyway. In any event, she said, she was glad to see Ann again and very willing to give her back her sick, weak cat—*if* she wanted her.

Ann thanked the woman, held back from saying anything that might have made her sad, hoping, she said, that things would go better for her from now on, and, wrapping her shawl carefully around Tinkle, they left.

Back in their apartment Ann began trying to help the very frail little Siamese, feeding her tiny bits of chicken or fish with her fingers, giving her water or milk with an eyedropper, staying near her much of the time, petting her, talking to her, saying her name over and over until she heard the first faint sound of a purr—an unmistakable sign that Tinkle wanted to live. And soon after that the cat began to revive. And was soon well.

"A miracle," said the vet. "But only with you. Your cat wants *you*. Only you."

But, Ann whispered into the sleek silk ear of the little cat lying on one of her favorite cushions—"we can't always have what we want in life, ma petite Tinkle." And dutifully, with a sigh, went back to her pre-moving tasks of putting names on boxes, etc....

That same afternoon Eva phoned to tell Ann about Clara. Clara and her two cats. Word of Clara had reached Eva through a mutual friend who lived in the same building as Clara did—she on the second, Clara on the top floor—and, as Clara was widely known throughout the neighborhood as someone who fed stray cats and kittens, Eva thought it might be worth a try.

Ann did too. Ann had actually heard about Clara before, had even met her once at a Christmas Bazaar to raise money for the local SPCA. But she had forgotten all about her till now. Now she remembered hearing of how warm Clara's heart was towards all animals, but especially toward cats. That she had a large apartment on the top floor of an old building not far from the market and near the beach. She picked up the phone.... A miracle she needed.

Invited to visit, when Ann arrived she saw that Clara's apartment was large and comfortable, even had a big outside patio which she had made into a lovely garden—a wonderful safe place for cats—and Tinkle loved plants and flowers.

"For how long?" asked Clara, her voluminous body settled back in a blue overstuffed armchair in her cool living room.

"I don't know. For an indefinite time of months and months. At the least," Ann replied, tears clearly shining in her eyes.

Ann reported to the FTRC (Friends of Tinkle Rescue Club)—inviting them to her apartment—though only Eva and Marjorie were able to come.

"See," she said. "If you do what is necessary, with hopes and prayers and with love from friends, then help does come. And in quite unexpected ways. Thank you."

Well, dear readers, after all the packing was finished and there was no more room for Tinkle to hide in the suitcases, then, as arranged, Ann brought Tinkle to Clara's. Up the seven floors in the old creaking cage elevator common in the older buildings of some French cities. Tinkle was in her small-sized carrier from which she could peak out—and all her things—toys, blanket, food, brush, leash and other special items in a basket in which Tinkle sometimes slept. At last Ann and Tinkle got out of the narrow elevator, walked to the door, and Ann rang the bell.

There! It was done. They had arrived at Clara's.

Let me not forget to tell you that it had also been decided that, before getting on the train, Ann would stay in Clara's apartment for a few days, in one of her extra rooms. The idea—actually the fruit of an emergency meeting of the FTRC with all members agreeing this was important, even essential—was to help Tinkle "adjust," as they say, to living in Clara's house. For cats, almost all cats, have a horror of change—and certainly little Tinkle did after 22 years of special routines, smells, sights, and the same much loved and trusted owner for all those years.

Well, what do you think, dear friends? Will Tinkle settle into Clara's *menage*? (French, if you please, for house or home.) Well, Tinkle will have to, because "After all, she is ONLY A CAT," as some people said, shaking their heads disapprovingly at all this fuss about "only a cat."

Thank goodness, Clara thought all cats very, very important and called her new Siamese guest "Miss Tinkle," and treated her with respect, which all cats appreciate. And especially Tinkle.

Oh yes, Clara. We must speak more about Clara—and her two cats, Pussy and Mitzy. Quite unusual cats. So much like their owner, Clara, with her rather particular ways. About what? Almost everything.

For one thing, Clara could never, would never, throw anything away, collecting things just like a—you've heard of packrats, haven't you? Well, Clara with her habit of saving everything still had every old newspaper she'd ever bought. Can you imagine it? Can you see this? —her house like a garret, full of boxes, packages, suitcases, bundles of newspapers in corners, on shelves.

Of course her cats didn't mind a bit because it's well known that most cats love the clutter of garrets, attics, and such places. There is so much to explore, you see—to be curious about—and Clara's cats were quite accustomed to this interesting atmosphere.

Now, about Clara's cats. Mitzy (a *French* name, you know), a black cat with green eyes, now eight years old, had been abandoned on the streets when hardly grown. A kitten herself, she had given birth to four other kittens. Clara had saved Mitzy the mother but, sadly, after the cold street and hunger, the kittens soon stopped breathing.

Well, Clara told this story about Mitzy daily (maybe as many as three times a day), until Tinkle felt like saying, "Enough! you've already told us that twice today!" Only Tinkle was too polite to say that, so she just sat and listened patiently.

What was really unfortunate was that now Mitzy had almost no teeth, at least no good ones, so she ate very poorly. Clara would explain—again and again—that it was Mitzy's babyhood as a stray that was to blame for her bad teeth. Which is why Clara would cut up Mitzy's food very fine—in teeny weeny bits. At the same tine, while standing at the kitchen counter cutting them up, Clara would promise, "Oh, soon I will take Mitzy to the vet to get her teeth fixed." Which Clara never did.

Someone else very important in Clara's household was Pussy. Pussy was perhaps the biggest cat in the world. Huge. Colored (as you can see from the drawing here) grey and white and striped. Pussy could easily have made three normal-sized cats. Grey and white Pussy was enormous because Pussy was a "gourmand"—which is not a "gourmet." You know the difference? Well, just to be sure, a gour*mand* loves to eat—ANYTHING—just to be eating. A gour*met* loves to eat only the most delicious and finest of foods.

Pussy, let me tell you, was a *gourmand*, and so of course, very, very fat and not very healthy.

But Pussy was always making believe that he was starving. Waddling like a duck trundling from side to side, he moved slowly to reach his plate of food. Crouching there, after a few morsels, he would have to get into a lying down position to eat because his hind legs wouldn't, couldn't support his oversized body. Then, after eating, he would flop over on his side or lie on his back, paws straight up in the air, waiting for the next meal.

When Clara put Tinkle's food on the floor, Tinkle would sit obediently before her big plate overflowing with sardines, chicken, lamb, steak, once even caviar, the most expensive food in the world. And Tinkle would try to eat a lot. But being small (and wise) she would soon give up.

Pussy would then get to work and clean up Tinkle's plate too. Which of course is why Pussy was just about the biggest cat in the world.

"Well," Clara would explain to Tinkle or whoever would listen, "Pussy being so big needs a lot of food." (*Au contraire*, French for "not at all" or "on the contrary," Clara had the cart before the horse, didn't she? Pussy was fat *because* he ate—and ate, and ate.) Poor, dear Pussy. It wasn't his fault. It was his dear owner Clara who fed him too much and too well, till Pussy had even developed a heart ailment from being so fat. Then Clara would give Pussy medicine for his heart, which beat "help, help, don't feed me so much." But to Clara it was a way of giving love to her animals—and she meant well.

Now Tinkle, let me tell you, understood very well what the situation was: that this was not a new home for herself *and* Ann. That, after all, *she,* Tinkle was the stranger in this house of Pussy and Mitzy's. And, of course, also Clara's. And that she, the newcomer, had to be very careful not to offend them in any way.

Above all, Tinkle wanted to be *good,* to fit herself into this strange new life in Clara's house and not be "trouble" or interfere with her two unusual cat companions. They *had* to become her friends. So Tinkle, almost holding her breath, tiptoed cautiously about—through the strange rooms and on the patio—watching Pussy and Mitzy, watching Clara, watching Ann—trying hard to cooperate. She could tell at once, being a cat and familiar of course with cat language, from the twitching of the cats' tails and ears, or by delicately touching noses (an inquisitive and sometimes friendly gesture)—if Pussy and Mitzy were in nice or nasty moods.

But Pussy could not be trusted. Every time Tinkle carefully walked by Pussy, he would try to sideswipe her. Why did Pussy do this? Was he angry? Maybe Tinkle knew, because Pussy kind of winked at Tinkle as his paw swung out lazily at Tinkle every time. Maybe *this* was Pussy's pet game. It did seem to be his only amusement.

Observing all this, Ann could not help but worry. Especially since Clara would lock all the cats up in her bedroom whenever she went out to shop—All three cats in a small room together with all her collections—strings, boxes, packages, old newspapers and magazines and photographs, and whatever.

Tinkle actually was afraid of being left alone in this interesting but somewhat frightening place, without Ann, her dear protector. Not that she ever blamed her—Only sometimes with her sweet, gentle, throaty cries, Tinkle would beg Ann—"Please, please, take me with you."

So even before Ann was to leave, some members of the Friends of Tinkle Rescue Club began to pay visits, as if to tell Tinkle—it's all right. We'll be watching over you..." Marie came, with a little catnip mouse, Eva came twice, and even the vet came to check over the new surroundings—and was quite astounded at Pussy's size.

But it was only when Ann came to sit quietly next to her on the terrace, or in her bedroom, that Tinkle felt almost calm—

"I love you, little Tinkle cat. I don't want to leave you, believe me," Ann said to her over and over, as she petted her on head and chest, and rubbed her behind her ears. Then Tinkle's very blue eyes would blink, or she would turn to give her own smooth pale fur a few licks, a way she had when she didn't know what to do next about a difficult situation.

And what a situation. Ann watched and waited. It seemed to her more often now that Mitzy and Tinkle would sniff noses, which meant, in cat language—"How do you do?" But there was no other *real* sign of friendship yet. No bumping of friendly foreheads, no lying close together in the same basket or chair....

Until. The very last day. The day of Ann's departure, the last precious hours when Tinkle and Ann would be together, something quite strange happened. One of the most surprising things Ann had ever seen a cat do. Something *very important*!

Dear readers, try to picture this. In one sudden unexpected move, Tinkle, with one high jump, sat down, first beside huge Pussy sitting on *his* broad cushioned chair—then, beside Mitzy, sitting on *her* high-backed pillowed chair. First, beside Pussy, then beside Mitzy—TRYING AS HARD AS SHE POSSIBLY COULD TO MAKE FRIENDS WITH BOTH OF THEM—even daring to nestle up close to Pussy, then to Mitzy—

Why? WHY? There could be no doubt to anyone watching and knowing the situation that this was Tinkle's very own and wise way of trying to become pleasant with these cats—Pussy and Mitzy—while Ann was away. For Tinkle truly believed she had to wait for Ann to come back to her from England.

Witness to this event, Ann marveled—"Amazing! She can think! How much she knows."

And then she hurried to the phone in the hallway to relay the latest news to the FTRC.

"No!" said Marjorie. "You aren't serious. I didn't realize she was such a deep one."

"Yes, Ann replied—"that's exactly what she did and, looking around the door, said, "she's still sitting there— right next to the big one. She can think. Actually think..."

Elizabeth laughed—"Of course they can. Even if they don't know that's what they're doing..."

After she'd hung up the phone, Ann sat there still in a daze—It was as if a door had swung open—her heart was filled with love and gratitude for this courageous little animal—who was not quite as dependent on her as she had thought.... Thank heavens.

Stooping quickly, she gave the little cat a warm caress. The taxi was already downstairs, her suitcases on the landing near the elevator. And so out Ann went with hand luggage and an extra coat over her arm. In her hurry—it was already late, she might even miss the train—she put all thought of Tinkle out of her mind—till later.

In her pillowed basket, Tinkle, left behind, buried her face in the fabric.

"Go away," She seemed to be saying to the world.

First, Tinkle wouldn't eat. After two days she swallowed a nibble, then another—to have the strength to wait patiently for Ann to return. She couldn't, wouldn't consider that Ann might not come back. Because in all the 22 years of her cat life Ann had always returned to Tinkle from any trip or separation. And Tinkle, being healthy and optimistic, had trust. Only this time she felt different in some way she couldn't of course understand—just feelings.

Very soon after this the alarm went out to Tinkle's friends as soon as Clara noticed how things were with Tinkle. For she too was now a member of the Friends of Tinkle Rescue Club, and a very important one.

Soon word was out.

"Tinkle isn't eating! Tinkle is sad!"

"Tinkle is crying!"

"With real tears," said Clara.

Then, one by one the Friends-of-Tinkle Rescue Club came to visit Tinkle in Clara's house, to bring her cheer if they could. And Tinkle was happy to see these old friends. They were like promises to be with Ann again.

The first friend to arrive was Marjorie.

"Fear not," was her greeting to Tinkle today. She had brought her special music box, sharing her musical passion now with Tinkle as she used to with Ann. Chopin, as always, entranced Tinkle, who sat close to the pretty little jeweled box, attentively still except for her ears twitching rhythmically.

"Don't worry," Marjorie said as she kissed Tinkle farewell. "I'll come again soon. And bring my violin."

By and by Eva came. Although she loved birds above all, and having her parrots and little birds in her home, naturally could not trust a cat there—not even Tinkle—still, Tinkle thought Eva the kindest of Ann's friends.

Perhaps because Eva had suffered so much in Germany in wartime she had learned how important it is to be kind. Well now Eva brought toys to Tinkle, carefully setting down a little bird cage, a red ball—and a tiny wriggling snake. A real one!

"Not all snakes are poisonous. In fact, very few are," Eva said reassuringly as Tinkle watched with some concern, but interest too. Carefully she put the little snake back in a grass-lined box with holes in it. Then Tinkle touched her little nose to Eva's which was a way she had of saying "thank you."

"Thank *you,*" said Eva, giving Tinkle's whiskers a tug—one habit which Tinkle didn't quite like.

Then Elizabeth came. Founder and President of the FTRC (the Friends of Tinkle Rescue Club), so caring of her own cat, Topolino, and of other animals as well, Elizabeth of course wanted to do all she could to help and save Tinkle. Listening to Elizabeth's high sweet voice, Tinkle almost forgot her misery and boredom.

"Well, my little Tinkle, what do you do with yourself all day long while waiting for your Ann?"

And Elizabeth went on telling Tinkle about Topolino, her own shiny chocolate-brown Abyssinian cat—(a very rare breed), who did not like his own name—*Topolino.*

"Because," he said, to those who could hear, "I am not a small mouse" (which is what Topolino means in Italian). And Elizabeth would reply, laughing "Oh Topie" (his pet name, which he preferred).

Thoughtfully, Elizabeth had brought a nice new brush as a gift, which pleased Tinkle, particularly when Elizabeth brushed where Tinkle couldn't get to herself with her little rough tongue—as Ann always had done.

"Now then," said Elizabeth in her soothing voice, "let me tell you a story. It's a true story of something that has happened at our house—in our lives..."

For Elizabeth, being quite a wise person with deep understanding of animals, having wondered and wondered how she might really be able to help Ann's little cat, thought—might it not give Tinkle more strength if she could hear of the misfortunes of other cats, and not think only of her own sadness?

So Tinkle, sitting as she often did with her front paws tucked under her chest, back ones properly folded also, on the upholstered footstool and Elizabeth in Mitzy's chair, prepared herself to listen as Elizabeth began to tell Tinkle a most interesting tale...about Topolino and Topolino's neighbor cat, Valentin.

"A large magnificent orange cat, Valentin longed, it seemed, so much to be Topolino's friend it was his only ambition in life. But alas, it was a friendship that Topolino spurned—who could tell why?

"Even though, mind you, strong Valentin would come each day across many dangerous slippery rooftops to arrive at Topolino and Elizabeth's high terrace. And there he would beg to be let inside the large window so he might sit with Topolino, always hoping Topolino might cry 'Valentin, I am so glad to see you! Welcome!'

"But no, regardless that I thought Valentin the bravest of cats—Topie (short for Topolino, remember) would have none of Valentin's friendship.

"Was it Topie's territory that he would not share with another cat? Even for short visits? Was it fear of losing his home? Whatever, there it was and nothing to be done.

"But Valentin had faith and didn't give up. He was a true, loving friend. So faithfully he kept coming across the sliding, slanting, dangerous roofs, meowing and mewing and trumpeting his arrival in his usual way, coming to visit his unwilling, but fascinating, friend, Topie.

"One dark rainy day, crossing and picking his way across the slippery slanty roofs—oh no, oh—Valentin slipped and fell down—down, down, down—the six floors of the apartment building, landing with a thud on the stone pavement below.

"An awful hush. There lying silent and unconscious in a little pool of blood was the big courageous cat. Everyone who saw him thought: 'This is the end of Valentin and his foolish visits to Topolino.' And all the neighbors whispered the bad news, shaking their heads sadly."

Elizabeth looked to see how Tinkle was reacting to this true tale—Yes, Tinkle's ears were forward, her eyes big and fixed and most interested in what Elizabeth was saying almost, as Elizabeth paused, she could imagine Tinkle saying—"Go on...go on!"

"But wonder of wonders—after a few weeks' stay in the animal hospital—where I took him with so little hope—the great-hearted and great-shouldered Valentin recovered from his broken hip! There were no other bones broken! Isn't that amazing?

"And one day, with a limp (all that remained to mark that awful fall)—one surprisingly bright day, Valentin's murmuring meow was heard again, announcing his arrival to visit Topolino. Valentin's heart was true, and, just as brave as you please, he had started again across the roofs to visit his beloved Topie—

"But, oh dear, oh dear—by then Topolino and I had moved to another street."

"So that's life, is it?" Tinkle asked unhappily.

"Broken hearts and gladness too," Elizabeth answered with her sweet smile as she said farewell to Tinkle. For she had to hurry home now to feed jealous Topie who would not eat unless Elizabeth was there to sit beside him and watch him eat.... Gracious, a difficult situation for Elizabeth, don't you think?

But what of Tinkle? The little cat sat alone very quietly for some time after Elizabeth had gone out the front door, and after she had heard her go down in the lift. The story of Topolino and Valentin had made her see life a little differently—Poor Valentin—and yet, he had not given up hope. Life was certainly not easy for anyone—or any cat, was it? It was up to her to show that she too, she, Tinkle, could be very brave and full of hope. For things did change...if one could only wait....

So Tinkle did try, very hard, to be happy. At least a little happy.

When letters came from Ann in England, Clara read them to Tinkle in the evenings when she came home from shopping or work.

The Friends loyally kept visiting—

Still, Tinkle could not help but grieve some, remembering in her body how Ann's soft hands felt on her head and back—how she looked walking in their apartment—or when she came in the door carrying groceries and, putting them down, would always come right to her to pick her up and call her name half-singing—"Tinkle, Tinkle—I'm back..." And so many other smells and sounds and feelings....

Then all Tinkle felt she could do was put her head between her paws and sleep—and sleep—so that sometimes a dream might even make her believe things were as they were...until she woke up.

The Friends of Tinkle Rescue Club met together—concerned. "What will happen to Tinkle? Will she eat enough to stay alive?"

And some said—knowing Tinkle's age—

"Tinkle is too old. She will die anyway. Maybe to save her all this grief and misery she *should* be put to sleep."

And it was decided to write this thought to Ann in England—

"Sleep?" Ann gasped in sorrow—"Put to sleep?" as she held the letter in her hand—"But Tinkle doesn't want to go *to sleep*—not really. She wants to play with leaves and branches and balls—to run around the terrace—to lie in the warm sun and watch the sea—to watch all interesting things going on around her—and smell them..."

For Ann felt strongly, and I'm sure some of you readers do too, and then wrote in her reply to the FTRC, "No, Tinkle is NOT OLD. She is healthy, strong, has a joyful, playful spirit and is wise beyond most humans' understanding. And cats—some cats—have been known to live a very long time. In fact, there are cats alive right now that are 28, 29—and at least one that was *over 36 years old* when she died—by being run over! (And this is a fact in the *Guinness Book of Records*.)"

So, dear friends and readers—the task is to hope and wait and help Tinkle survive in Clara's house—without Ann, yes, but still expecting Ann to come back. And who knows?

Then, there is the mystery of love—always. Tinkle does not dislike Clara—whom she is only beginning to know. Mitzy sometimes almost plays with her—Elizabeth says she will bring Topolino to visit soon....

In the meantime, please, all of you do your best in your thoughts and hearts to help dear Tinkle be well—to have the wish to live, to be strong and full of trust and hope in all good things—even perhaps to be with Ann again!

Goodbye, dear friends and readers. Please—will you join the Friends of Tinkle Rescue Club?

About the Author—

Each little flower that opens,
Each little bird that sings,
He made their glowing colours,
He made their tiny wings...

A moth, a squirrel, a weed pushing up between the cracks of concrete or stone—all such events were a source of delight and wonder for Ann Eve Moss, my mother, author of this small book. A belief in the possibility of miracles and above all in the miraculous quality in all life from the tiniest particles of matter to the huge entities we call planets gave her life a special sense of adventure, meaning, and joy. And this remained true all her life, even into her eighties.

Many people sought Ann's friendship in her later years because they had never known anyone at that age as excited by life as she was, a person still so occupied and growing. They found it inspiring to know someone who, well into her ninth decade, painted, wrote poems, travelled, went to Elderhostels (where she studied biology, physics, geology, jazz, religion, politics—and fell in love with Thomas

Jefferson), grew healthy plants and tomatoes on her terrace, and was busy rewriting two long novels.

Born in New York City in 1903, Ann (then Anna Dressler), was taking dancing and singing lessons while in high school and was on the Broadway stage by the time she graduated, as much because of her theatrical talents as for her natural remarkable beauty. Every life has its special gifts. Ann had many, and quite soon in life she decided to try to realize these as fully as she could, and not waste time in developing her capacities. Her intention came partly from a wish to help with the practical needs of home, but also from a deeply rooted sense of responsibility to others, to the world and to that which she could not fully define when young, but later in life called "the divine push of evolution."

Ann did not speak of her childhood enough to satisfy my curiosity. She was quite reserved about those early years in New York City. But one episode she told me several times when I was a child made a lasting impression. This was the story of how, as a girl of six or seven, growing up in a crowded poor neighborhood of downtown Manhattan in the first decades of this century, having few toys of her own, she often found kittens or puppies to play with. These were abundant—mostly homeless creatures. Sometimes they were welcomed to live in the back rooms of restaurants, groceries or bars, and were fed surplus milk or scraps in exchange for their services as mousers or ratters.

On this particular day, as my mother recounted, she was (she remembered) feeling unhappy. Her mother or father had perhaps scolded her. Their lives were not easy. Her mother took in sewing and raised four children on a very small household allowance given to her by her husband, who sold toys on the street. While playing in an alleyway near her apartment building with a small kitten (she recalls it as being a grey-striped little tabby), she popped it into a brown paper bag, thinking it might be fun to whirl it about, and twirled it faster and faster in front of her. Tired of this activity, she laid the bag down, but on opening it saw the kitten lying very still. When it did not respond to prodding or shaking, she began to cry.

Until her last years this memory brought tears to her eyes and feelings of guilt. It was hard even then for her to forgive herself. And throughout my childhood, during all the years we lived in the city, no stray cat was ever passed without my mother giving it some food, even if she had to run several blocks to find a grocery. Often a phone call was made to friends, or some animal organization, in an effort to find a home. Several times we were the ones to "foster" some unwanted creature—sometimes for years.

Ann observed nature carefully. She watched how birds and animals had such a struggle to survive, to find food in winter, coolness and water in the heat of summer. She admired how carefully and beautifully they lived their individual lives, keeping themselves clean, providing for their young. The exquisite architecture of birds' nests, the cooperation of ants in their tasks, the harmonious movements of fish and sea-beings in their watery element—such phenomena never ceased to amaze her and sustain her in her belief in the goodness of life. Despite the harshness she often encountered in the rough-and-tumble battle for success or even survival in New York, there was never any doubt in her mind that "Each little flower that opens, each little bird that sings..." (from her favorite hymn, "All Things Bright and Beautiful") were proof of a Creation that is essentially good.

This attitude and perspective probably explains why, even when her physical beauty and theatrical talents were publicly praised and acknowledged—with her photo often appearing in newspapers and magazines, jobs in top shows and musicals offered her, adulation afforded her as a "Zeigfeld Girl"—she never exhibited or seemed to feel much pride or vanity in herself, interested as she was in the larger dimensions of things, engaged even then in studying philosophy, politics, and the social issues of the day.

Not that she was any more protected from the dangers of the New York theatrical scene of that time (1920-1935 or so) than any other attractive young woman. It was hard to tell, for instance, whether the men who courted her and asked her out did so for who she was in herself, or, as she more often suspected, from a wish to be seen with her because of her "looks" and fame—or for even less honourable reasons. It was well known that one way to success was the open bedroom door.

But Ann was determined from her teens and earlier, partly from a moral set of values at home, but also out of her own convictions, not to become part of that sort of casual life-style in the theatrical milieu. So that, when in 1922—she was not yet twenty years old—she met Harry Moss, a theatrical agent managing many of the top jazz and popular orchestras of the time, a person who, like herself, was interested in greater issues, gifted in music and writing as well, they were married the same year—not long before she joined the Zeigfeld Follies.

As a Follies Girl, Ann appeared in "The Three Musketeers," "Garrick Gaities," "Funny Face," and others. She was front-page news in 1928 when she refused to audition in the nude for producer Earl Carroll's new show, "Fioretta," and then went so far as to bring charges against him before Actors Equity, the theatrical union, for making exposure a condition for being hired. She won the case in January, 1929.

Her willingness to speak out against injustice was unusual for women of that decade. In a review of her life just after her death in 1988, a prominent newspaper interpreted it as a "feminist protest." But it was far more than that, in the light of her deepest concerns as expressed in many of her writings: that trust, love, creativity, joy not be destroyed or damaged by economic, political, or sexual forms of exploitation. The producer, in her view, by treating her body as a product to inspire lust was degrading both the body and sexuality. Secondarily, she saw that it was his position as a successful male that gave him, he thought, the right to do so.

Aware of the fragility of true sexuality, Ann saw, far ahead of the present phenomena of broken homes and the single parent, how a breakdown in the upholding of the best feminine and masculine values (equal in her eyes) could destroy the possibility of healthy, loving relationships, and therefore society itself would suffer—as it has—with an increase in crime, mental illness, addictions.

She was almost twenty-four years old when I, her first and only child, was born. Although she did not give up her career for very long, she earnestly endeavored to be a good mother, nursing her child for nine months, quite unusual at a time when the style was not to nurse children at all if possible, particularly if you were a beautiful woman in the New York theater.

For the remainder of the 1920s and well into the '30s, Ann was busy appearing in N.Y. musicals and "shows," going "on the road" with them for weeks at a time all over the country. A part in "Music In the Air," produced and written by Jerome Kerns, gave her an opportunity to sing songs she especially loved, and never stopped singing—A few days before her death she sang a few bars of her favorite "I've told every little star..." There were also modelling jobs, nightclub performances, radio soap operas...These were often arranged by her husband, Harry, who had become an important agent in the N.Y. musical scene, and later an executive for MCA (Music Corporation of America).

Signs of another big war across the Atlantic were being received in the United States as Adolf Hitler began to mobilize his troops and equipment to threaten neighboring nations. Alarmed at the memory of World War I, not so far in the past, my mother invested time and thought trying to comprehend the deeper causes of conflict, reading, listening to as many viewpoints as she could. Some neighbors and relatives, even friends, thought this behavior rather odd. But then Ann was already considered offbeat for her interest in diet and healthfoods, relatively new at that time. But despite the ridicule she often received because of her belief that there must be some important connection between health and what we eat, she made it a point to always look at the labels when she shopped—and to buy foods as natural and unprocessed as possible. Unfortunately, at that time, neither I nor my father were particularly appreciative of her efforts.

Languages came easy to my mother. Especially French, because France was always—as far back as I can remember—the second country of her heart. Despite rumors of an imminent outbreak of hostilities on the European continent, she decided to make her first voyage to France in the early summer of 1938. It proved indeed to be the last chance to go there for many years. Almost 11, I was to be her travelling companion. We boarded the S.S. Normandie in June, with an unusual assortment of shipmates, including an entire battalion of soldiers called the Loyalist Brigade, on their way to fight against General Franco, a later ally of Hitler's.

Somehow we put all the war-consciousness aside and had a joyful summer. In France, Ann discovered an almost inexplicable sense of familiarity and at-homeness which was to remain part of her all her life, prompting several future trips, and, last, her eight-year stay on the French Riviera—with Tinkle, of course.

On our return to the United States, Ann launched fully into her writing, completing a draft of her first novel, later titled *Catha's Sister*, based on her experiences on the stage, and seeing herself in a new and more mature perspective.

> Racing up three flights of stairs at breakneck speed for a fast change, intuitively keeping her face calm, no facial mannerisms to impair it—in the brilliantly lit glass a revived childlike face with planes that artists, already sketching in their minds, remarked with delight—a face that saw clearly...

But writing is seldom lucrative. She began seeking other ways of earning a living, a need becoming more serious as her marriage of over twenty years was coming to an end. Now her skills in shorthand and typing in high school proved of much value, giving her the support she needed as a single woman in an expensive city.

Although Ann held a few long-term secretarial jobs with large established companies, she discovered that what suited her best as a creative artist needing time, independence, and flexibility was what is called a "free-lance secretary." This part-time work gave her the time to write when she wanted to, travel occasionally, and still work enough hours to pay basic expenses. This was, of course, before inflation had seriously set in, and even New York City apartments, then rent-controlled, were not exorbitant. In time she moved to a large studio apartment on West 57th Street that suited her needs and tastes, and stayed there for many years.

She found she could, more or less, pick and choose her jobs and clientele as a skilled secretary with a high degree of intelligence who was also comfortable with languages. Among her employers were Erich Fromm, Rollo May, Robert Jay Lifton (all in the field of psychology); millionaires—Michael Grace (of the Grace Steamship Line) and Countess Mona Bismark, for whom she later worked full time as administrative secretary on her estate in Capri, Italy; government officials, industrial tycoons, and, later in Nice—for Paul Pierre Matisse, son of the painter, Henri Matisse. She gleaned an inside look at the world in which she lived through this work, which later became the foundation of her second novel, a long and serious book (still unpublished) which, after my father's death in 1948, she titled *A Widow's Odyssey*.

In this book, too lengthy to interest any publisher she contacted, she allowed herself the luxury of expounding some of her ideas about "cut-throat economic policies that were inflicting grim injuries on the natural world, tearing the sensitive delicate fabric that joins all life"—and questioning how real love can exist in a "distorting society that spared no one." Some of the characters in the novel were obviously modelled on past employers—and as she rephrased their words there was a strong quality of authenticity—though at times she used them to get across her own ideas—admitting to me, and in her private journals, that she should have been a preacher. She was always quick to laugh at her own, sometimes too opinionated way of putting things, but saw her thoughts as seeds—for the future—some already, she thought, bearing fruit.

At last she felt she could afford to transfer her work and life to France, a long cherished hope. In the mid-1960s she moved to Nice, a fairly large town-city on the Mediterranean coast, taking the little Siamese, Tinkle—her cat companion, of many years—with her. Easily she merged her American life with that of French neighbors and friends—and felt very comfortable setting out mornings for the market-place with her string bag, meeting friends for coffee, walking along the sea. It was a happy time, and her writing flowed easily.

On leaving France, however, to go to England (where her daughter was indeed living and in need of her), it was still a hope that she might return to Nice before too long—to Tinkle and good friends. Instead, circumstances required she return, in August of 1975, to the United States.

Tinkle had, by then, adjusted somewhat to her routines with Clara, Pussy, and Mitzy with the help of members of the Friends of Tinkle Rescue Club, including the caring Vet. All humans in this group agreed that the air trip to the United States could be very hard on an almost-23-year-old cat. Ann agreed too, though not without episodes of pain and doubt—which drove her on to write about her problem.

Kay Beesley, founder of the "Cats Protection League and Tailwavers Society," replied to her query: "...I think if it were me, I would let the cat stay put and not give her the upheaval of such a long journey. Twenty-three is a marvelous age and would be over 140 in human years—so the lady must have taken good care of her—...It would be nice if you would adopt a nice unwanted cat wherever you settle..."

Which she did. On settling in Greensboro, North Carolina, Ann adopted a cat that had been destined for an over-full shelter and probably "euthanasia." This all-black cheerful female she named Nera was very different from Tinkle, of course. But she went on having dreams about Tinkle. One dream seemed to her to prophesy Tinkle's death. With little notice to anyone, one morning she boarded a plane for Nice, determined to see her long-time feline friend one more time.

Satisfied that Tinkle was cared for as well as possible, and that she probably would not survive a trip to the States, Ann said a final farewell, and resumed her life in Greensboro, knowing she had done all she could to conclude well this relationship of almost a quarter century. Grateful for the joy it had given her, strengthened in her feeling that animals and humans are meant to live together in harmony and help each other, she attributed much of her own youthfulness and health to the quickening of spirit that resulted from caring for Tinkle and learning so much from her ways—above all, learning how to play, really *play* into one's old age. It was one of Ann's hopes that in retirement settlements, nursing homes, and other communities for elderly as well as in private lives, older people would let themselves heal and grow younger through association with animals—which could also give millions of animals a chance to live...

A few weeks later, after Ann's return from France, Tinkle did let go of her little life peacefully and swiftly. That was the beginning of this book, *The Friends of Tinkle Rescue Club,* which gave Ann a chance to relive some moments with Tinkle—and immortalize one little old-young Siamese.

Ann left the then small city of Greensboro, North Carolina, in 1981 to follow her daughter—myself—who had taken a job teaching in an alternative country school in Virginia. The house she bought in the town of Floyd had a nice yard and a protective garage in which several stray cats found a happy haven. A few years later, past the age of 80, Ann decided it might be best to join a retirement settlement; the decision for one so independent of temperament was not an easy one. Yet once she had actually moved into a small apartment in Warm Hearth Village, near the mountain university town of Blacksburg—not so far from friends in Floyd or North Carolina—she began a life that soon became very full. Her involvement in church and local affairs, new friends, and daily time for writing, Yoga, swimming, exercise, studies, and a through awareness of world events amazed everyone who knew her, as did the quality of her constant readings—Emerson, Thoreau, Tolstoy, C.S. Lewis, Rupert Sheldrake...and of course Albert Schweitzer whose "reverence for life" confirmed her deepest feelings and hopes.

Here, too, animals—as if sensing a spirit to whom they were of value—found her. Birds thronged her balcony, on which there were always seeds for them. Squirrels ran up her wall to compete for the seeds.

She tried to keep the peace—and enough seeds. Eventually a gentle young white stray cat that wandered the gardens discovered Ann. Ann's visits to her with food became a twice-a-day ritual, and she named the adopted stray "Angel"—appropriate until Angel appeared with a litter of four.

On a Christmas morning in 1987 Ann received a pink Amaryllis not yet fully in bloom. Enthralled by "such mysterious and superb grandeur," she wrote about it at length as it opened...

> Erect, lily-like, these five huge opening belles on one thin stalk, arched all around in a magnificent circle of grandeur. For what? For me? I feel like bending my knees before the altar of such a flower...

A few days before going to the hospital for what proved to be an incurable situation, she was relieved and pleased to find excellent homes for all the kittens, and a very special one for Angel, the mother cat. A good friend also took her fish, Goldie and—another Angel.

All proceeds
from the sale of this book
over costs
are to go to relieving
some of the vast suffering of animals
through various animal organizations.

A quiet moment of trust between Ann and a stray cat—taken in Blacksburg, Virginia, in the 1980s.

Ann Eve Dressler at sixteen, studying drama and dance in New York City (1919).

At eighteen, during a modern dance class.

Ann and her husband Harry with their new baby, Marilyn Allyn.

Ann and Harry Moss on their honeymoon in Atlantic City, New Jersey, in 1923.

Ann Eve Gaynor (a theatrical name) as a Zeigfeld Follies Showgirl in 1928. She later refused to pose for photos of this type.

Portrait-photo of Ann circa 1925, when she was much in demand as a model for many major magazines, newspapers, and modeling agencies.

Taken in the early 1930s. As Ann's inging career grew and her appearances on tage and in nightclubs became more frequent, his photo was used for advance publicity and on tours.

Tinkle at three months of age.

Tinkle at 20, resting on Ann's kitchen table in their apartment in Nice, France, about 1973.

Tinkle having her supper between Pussy and Mitzy at Clara's.

Pussy—who "could easily have made three normal-sized cats"—waiting between meals.

Ann in Nice, France (1974),
on the terrace of Clara Aubert's apartment.

Ann at a restaurant in Europe during the time that she worked as a free-lance secretary for millionaires and leading figures in the arts and sciences.

Spring, 1979, in Greensboro, North Carolina. In the garden of a friend (Emily Huntley), Ann makes a new cat-friend.